NAVIGATING THE INTERREGNUM

CROSSING INTO OPEN SEA

David Kantor

Urantia Book Films Press
Lakewood, Colorado

ISBN 979-8-234-01646-1
Published In association with
Rocky Mountain Spiritual Fellowship
Boulder, Colorado

Disclaimer
This book makes reference to concepts from The Urantia Book. The interpretations, commentary, and views expressed herein are solely those of the author and do not necessarily represent the views of The Urantia Book Fellowship or any other reader organization or publisher.

First Edition 2026
Cover Art: David Kantor, Midjourney
Urantia Book Films Press
Lakewood, Colorado, USA
admin@UrantiaBookFilms.org

A sower went forth to sow . . .

Dedicated to those farseeing and forward-looking
men and women of spiritual insight who will dare
to construct a new and appealing philosophy of living
out of the enlarged and exquisitely integrated modern
concepts of cosmic truth, universe beauty, and divine goodness.

"The object of hope is a future good;
difficult, but possible to obtain."

Thomas Aquinas

Table of Contents

Preface

Topics of conversation among readers of *The Urantia Book* and in the councils of Urantia institutions invariably include: "How do we get people interested in reading *The Urantia Book?"* "How do we make the world aware of the revelation?" "How should we approach dissemination?"

These conversations often develop along ideological lines, shaped by personal opinions and second-hand, apocryphal claims. But these questions cannot be addressed rationally without first understanding the cultural environment into which the book will be introduced—and how that environment, and its constraints, are likely to change over time.

Just as you would never take off into the mountains on a backpacking trip without first checking the weather, neither should we implement a dissemination plan without first checking which way the winds of change are blowing, and what the clouds on the horizon portend.

Navigating the Interregnum attempts to derive an informed dissemination ethic from the text of *The Urantia Book* itself, read alongside the diagnoses of social theorists, historians, and visionary Christian leaders. The urgency of this inquiry is sharpened by recognition that we are entering an interregnum—a period of prolonged instability that may not quickly resolve.

My hope is that *Navigating the Interregnum* will catalyze sober reflection on what wise dissemination might mean under our actual conditions. A tightrope must be walked. On the one hand, the world needs this revelation to reorient itself. On the other hand, if it is perceived as a new religious movement or sect, that perception will retard acceptance and may extend the suffering and disorientation intrinsic to an interregnum.

The interregnum is a crisis of lost spiritual values. The primary insight of *Navigating the Interregnum* is that only persons can generate spiritual values. Social structures and institutions instantiate those values but are powerless to create new ones. The horizon for effective recovery and renewal gains is the life of the individual. This is the challenge for revelation dissemination—a work of global consequence during a period of unprecedented planetary transition.

Orientation to Key Terms

Here are several key terms used throughout this book. Being familiar with them will greatly enhance understanding and smooth your way through the text.

Ultimacy

Ultimacy, as used herein, names the highest-order object of loyalty and worship that quietly governs what a person treats as finally real, finally authoritative, and finally worth living for. It is the experienced anchor of value that orders conscience, priorities, and sacrifice, and that supplies a destiny horizon by which suffering, duty, and hope become intelligible. It is the highest value by which all the rest of one's values are calibrated.

In an interregnum, ultimacy is not absent but displaced: The need for an organizing center migrates from a vertical spiritual orientation to lower, horizontal substitutes (ideology, nation, tribe, self, system), which can energize life for a time but cannot bear the weight of final meaning without deforming both persons and institutions.

As used in this book, ultimacy is employed in an existential and functional sense (as in Paul Tillich's "ultimate concern"). It denotes what is treated as finally authoritative and worthy of devotion—not in the Urantia Papers' more technical cosmological sense of the Ultimate as a superfinite or absonite level of reality and deity expression.

Vertical Axis

The vertical axis, as used herein, is the interior dimension of human life in which a person relates to ultimacy—the realm of God-consciousness, worship, moral sovereignty, and destiny-orientation. It names the direction of experience in which conscience becomes more than social conditioning, value becomes more than preference, and the self becomes answerable to something higher than appetite, tribe, or historical fashion.

To come into vertical alignment is not to withdraw from the world; it is to regain an inward anchoring—a lived reference point of divine reality—by which motives are purified, priorities are reordered, spiritual values are established, and the will can be progressively unified around the doing of the will of God.

The teachings of Jesus illustrate the horizontal and vertical axis model. Jesus gave us but two spiritual practices: Worship and service. In worship we cultivate the vertical and in service we serve the horizontal.

Horizontal Axis

The horizontal axis, as we are using it, is the outward, time-bound dimension of life in which persons and communities interpret experience, coordinate action, and build workable systems—the realm of meaning-making, social roles, institutions, culture, and history. It is where spiritual values generated on the vertical axis must be *translated* into intelligible narratives, stable practices, and concrete forms of love and service amid complexity and change.

When the horizontal axis is well-ordered, it provides an inhabitable universe frame which gives proportion to suffering, steadiness to duty, and coherence to hope. When it is disordered—especially under displaced ultimacy—it becomes crowded with substitute meanings and overloaded systems that try to do the work of ultimacy and eventually fracture under the weight.

Think of the horizontal axis as your participation in the evolution of the Supreme in the domain of time.

Spiritual

The term "spiritual" used in this text implies persons and their relationships with each other and with God, mediated by their Thought Adjusters.

The Urantia Papers

The Urantia Papers are published as *The Urantia Book.* I use *Urantia Papers* herein to avoid formatting inconveniences. It's also noteworthy that the revelators use the term in referring to the revelation.

What is an Interregnum?

In Paper 195, *interregnum* names a prolonged civilizational interval in which vertical-axis anchorage has weakened or collapsed, while no new, spiritually adequate integration has yet emerged to replace it. It is not a brief transition or a temporary confusion; it is an extended orientation vacuum affecting meaning, value-order, and social coordination at every level of Western life. In this sense, an interregnum is not merely "decline."

It is a period *between reigns:* between the effective cultural sovereignty of traditional Christianity and a future form of spiritually vital religion capable of guiding a planetary civilization.

Its defining feature is that meaning, loyalty, and moral coordination are no longer secured by a socially shared vertical axis of ultimacy, even as social complexity accelerates and the demands of collective life become more intricate. Societies in such a condition become unstable—not because horizontal institutions disappear, but because those institutions are pressed to supply what they cannot: moral authority, inward integration, and a destiny horizon.

When the vertical axis thins, the horizontal field also loses intelligibility. People inherit fragments of competing stories—identity, ideology, technique, consumption—none of which can carry the whole weight of suffering, duty, hope, and long-horizon responsibility. The result is not simply disagreement; it is a loss of shared interpretive architecture—a collapse in the meanings that make public life navigable and moral action coherent.

What emerges, therefore, is not merely chaos but a brittle order—one increasingly tempted to manufacture coherence through surveillance, regulation, procedural overload, and emergency powers, because organic integration has weakened. This is the phenomenology of the interregnum: a civilization struggling to function successfully on the horizontal axis, yet inwardly disoriented—trying to stabilize coordination without restoring ultimacy.

Cultural Symptoms

Following is a list of signature dynamics we could expect to see in an interregnum as described in Paper 195: a weakened vertical axis of ultimacy with horizontal systems competing to supply meaning, cohesion, and moral authority.

1. Axial Weakening and Displacement of Ultimacy

From a Urantia Papers perspective, the interregnum begins as shared sources of vertical integration—God-consciousness, moral authority, and a destiny horizon—lose cultural sovereignty. Society does not become neutral. It becomes procedural and contested, because the deepest grounds of loyalty are no longer commonly held. Mechanisms are asked to do the work once carried by living meanings.

Reference: Paper 99; Paper 81

2. Horizontal Substitution Pressure

Human beings cannot live without an organizing center of meaning. If the true transcendent center is eclipsed, the need remains and migrates. Instrumental structures—the state, ideology, movements, media—are loaded with identity and destiny expectations they metaphysically cannot support. This produces instability rather than integration.

Reference: Paper 1; Paper 5

3. Ultimacy Displaced into Politics and Identity

Politics shifts from coordination to existential drama: Who belongs, what is sacred, what must be defeated. Opponents are moralized, disagreement becomes threat, and the state-sphere absorbs meanings it cannot legitimately hold for any length of time.

Reference: Paper 195

4. Delegitimation and Migration of Authority

When inner moral integration thins, social trust becomes brittle and authority shifts to whatever feels coherent or protective. Institutional credibility turns volatile because legitimacy is no longer nourished by shared moral culture. In Urantia Papers terms, trust is downstream from spiritual vitality: Stable social trust depends on upstream sources of spiritual values, the moral-spiritual quality of persons, and the shared moral culture they generate.

Reference: Paper 99; Paper 71

5. Epistemic Splintering and Multiple Realities

The Urantia Papers distinguish facts, meanings, and values, and expect conflict when they are severed from spiritual anchorage. As shared moral horizons erode, shared standards of proof erode with them. Communities crystallize around incompatible priors. Persuasion loses traction because interpretive universe frames no longer overlap.

Reference: Paper 103; Paper 101

6. Privatization of Meaning and "Spiritual Shopping"

Spiritual hunger persists even as inherited religious authority weakens, producing individualized and unchurched forms of seeking. The Urantia Papers affirm personal religious experience as primary yet warn that religion must eventually find stable social expression. The distortion is thin transcendence: many private meanings, few shared obligations.

Reference: Paper 100; Paper 99

7. Loneliness, Atomization, and Weakened Intermediate Communities

Family and intermediate associations are civilizational load-bearers. Their erosion destabilizes both personality development and social order. Interregna generate the feeling of being unheld—relational precarity that heightens susceptibility to manipulative belonging-offers and totalizing narratives.

Reference: Paper 84; Paper 81

8. Teleological Collapse and Replacement Ends

When a credible future-bearing teleology fades, cultures default to immediate substitutes—comfort, stimulation, status, optimization. Time horizons shorten, values thin, and anxiety rises because suffering loses intelligible placement within growth. The Urantia Papers treat destiny orientation as a primary stabilizer of life.

Reference: Paper 5; Paper 100

9. Coercive Compensation

Law and government are necessary instruments but inadequate substitutes for inner moral integration. As fragmentation deepens, institutions are pressured to manufacture coherence through expanding external control. External enforcement replaces internal assent, increasing the risk of authoritarian drift.

Reference: Paper 71; Paper 70

10. Reaction Formations and New Traditionalisms

Prolonged uncertainty produces a longing for enclosure—strong leaders, rigid boundaries, restored hierarchies promising relief from ambiguity. The Urantia Papers caution that such compressive solutions often trade spiritual growth for enforced order, substituting fear or identity absolutism for genuine fraternity.

Reference: Paper 195; Paper 71

11. The Interregnum Tension: Suspicion of Ultimacy, Hunger for Ultimacy

The interregnum expresses a paradox: skepticism toward grand meanings coupled with heightened vulnerability to emotionally satisfying ones. When the true center is distrusted, hunger remains, producing oscillation between cynicism and credulity. In Urantia Paper terms, this signals a failure to re-anchor meaning in living spiritual experience and its social fruits.

Reference: Paper 195; Paper 101

Causes of the Deepening Interregnum

Secularism

In Urantia Papers terms, secularism is not merely "less religion"; it is the cultural removal of a unifying spiritual horizon that once integrated meanings, loyalties, and long-range purpose. When that integrator is withdrawn without a living substitute, society does not become more rational; it becomes more procedural, more contested, and more vulnerable to surrogate ultimacies. The result is an authority vacuum at the level of meaning, with horizontal systems pressured to carry what only inner spiritual orientation can sustain.

Reference: Paper 99; Paper 195

Materialism

The Urantia Papers treat materialism as a flattening of reality in which value is reduced to mechanism, utility, and measurable satisfactions. This does not merely change metaphysics; it reshapes conscience by shrinking the range of meanings that can command sacrifice and discipline. When the real is defined as only what can be managed, the good becomes what can be optimized, and the soul's hunger is displaced into restless acquisition and stimulation.

Reference: Paper 2; Paper 195

Nationalism

The Urantia Papers point out that nationalism can supply near-term cohesion by tightening identity within the local group, but it does so at the cost of blocking the higher moral demand of planetary solidarity. It strengthens the tribe while weakening the species, treating loyalty as a zero-sum possession rather than an expandable capacity. The long arc of civilization, in Urantia Papers terms, points toward wider circles of fraternity.

Nationalism reverses that trajectory by absolutizing the partial.

Reference: Paper 71; Paper 195

War and Power Politics

The Urantia Papers frame war and power politics as symptoms of misdirected ultimacy—attempts to secure safety and meaning by coercion when inner integration is lacking. When identity and destiny are assigned to the state-sphere, conflict becomes existential and opponents become moral threats, making violence easier to justify. In that condition, politics stops being coordination and becomes a theater of salvation by force, with predictable cycles of fear, retaliation, and hardening control.

Reference: Paper 70; Paper 71; Paper 195

Decline of Institutional Religion

The Urantia Papers distinguish personal religious experience from institutional forms, but it also treats institutions as socially necessary carriers of shared meanings when they are alive and ethically credible. When institutional religion declines, the culture often loses a stabilizing transmitter of moral consensus and destiny-horizon, creating an authority vacuum at the level of ultimacy. Spiritual hunger may persist, but shared discipline and shared language thin, and the social fruits of religion become harder to sustain at scale.

Reference: Paper 99; Paper 100; Paper 195

Family breakdown

The Urantia Papers highlight the family as the primary school of loyalty, self-mastery, and value transmission. When it breaks down, society loses its most basic mechanism for stabilizing character and conscience. The damage is not only emotional; it is civilizational, because habits of fidelity, service, and responsibility not formed early in life are formed later with difficulty. When families fragment, the culture becomes more atomized, and the burden of moral formation shifts to weaker, thinner substitutes.

Reference: Paper 84; Paper 99

Rapid modernization

The Urantia Papers treat civilization as an evolutionary growth in which institutions and moral culture must keep pace with expanding knowledge and technical power. Rapid modernization accelerates dislocation faster than adaptation; old frameworks weaken before new ones can mature, and technical capacity outruns ethical integration. The result is cultural vertigo—more options, more speed, more stimulation, but fewer shared meanings capable of guiding that power toward constructive ends.

Reference: Paper 81; Paper 99; Paper 195

Cultural fragmentation

Cultural fragmentation is what happens when a shared moral horizon no longer holds: Groups inhabit incompatible narratives, vocabularies, and authorities, making consensus feel coercive and compromise feel like betrayal. Without a common spiritual and moral center, plurality becomes not creative diversity but centrifugal division, and society struggles to coordinate around any enduring good. In that condition, the hunger for integration often returns in distorted forms—tribal absolutisms and coercive compensations.

Reference: Paper 99; Paper 101; Paper 195

None of these elements alone define the interregnum—their convergence does. Paper 195 mentions the interregnum almost as a passing comment in a context of warning about premature levels of attainment. But scattered elsewhere in the Papers we find multiple references to both symptoms and causes of an interregnum. A succinct definition of the interregnum is provided in 99:4.6.

When secularism tacitly or openly "revolts against God" and attempts to occupy the place of religion as the integrator of meaning and morals, the Urantia Papers frame the consequence as social unrest, animosity, war, and the danger of world-wide disaster. This is not rhetorical.

In that same logic, the liberation from ecclesiastical totalitarianism can be followed by a new captivity: the rise of a tyrannical political state and forms of political–economic enslavement, described as the offspring of materialism and philosophic secularism.

The Urantia Papers note additional pressures which are treated as accelerants of vulnerability. Nationalism is a paradoxical force—at once a chief cohesive power resisting disintegration and noted in Paper 195 as "the chief barrier to world peace." War-weariness and nationalistic militarism is called out in Paper 70 and appear as the harvest of misdirected loyalties and the lust for power. Christianity is criticized in Paper 195 for lowering its ideals under the pressures of greed, war-madness, and domination.

Paper 99 adds a cluster of destabilizing conditions: the sudden mixing of cultures, intermingling of creeds, diminishing ecclesiastical authority, urbanization, mechanization, and rapid economic and social change—all of which intensify moral crosscurrents and strain the capacity of institutional religion to lead reconstruction, since it is often entangled with the very social orders undergoing upheaval.

Paper 195 sharpens the critique further by warning of religion's drift into formalism, over organization, and intellectualism—"secondhand" religion replacing firsthand spiritual experience—precisely when living spiritual vitality is most needed.

Paper 84 decries a rising tide of self-gratification and "pleasure mania" is described as endangering marriage, accelerating the decadence of family life, and placing at risk the very mechanism by which civilization transmits spiritual values across generations.

It is precisely the convergence of the above factors which Paper 195 gathers under the term *interregnum.* These factors are not merely "contributing causes"; they *constitute an interregnum* when they occur together.

Paper 195 is therefore not merely lamenting decline. It is diagnosing a dangerous interval of planetary struggle. Old forms have been outgrown, but the next integration has not yet been lived into existence. The implicit warning is that a civilization cannot remain indefinitely without a shared horizon of meaning; the hope is that the next integration will be personal-first—then cultural—rather than the mistake of embracing institutional primacy.

Social Environmental Vectors as Amplifiers

The following section lists "vector pressures"—processes which are evolving in specific directions toward problematic outcomes. These are offered not as predictions, timelines, or a single causal chain, but as high-leverage forces that commonly converge when a civilization is living through an interregnum. The vectors are organized into two families.

The first eight are diagnostic interregnum vectors: interior and cultural symptoms that indicate a weakened vertical axis of ultimacy—fractures in meaning, trust, personhood, freedom, knowledge, moral life, institutions, and the affective/aesthetic atmosphere.

The second eight are intensifiers, external stressors—ecological, biomedical, and fiscal-system pressures—that do not "cause" the interregnum, yet can intensify it by raising baseline strain and tightening the feedback loops of fear, rivalry, and coercive compensation.

The relationship between these families is structural rather than simplistic. Environmental deterioration does not automatically produce an ultimacy crisis, and an ultimacy crisis does not automatically generate each environmental difficulty. But when ultimacy is displaced from a vertical horizon—truth, beauty, goodness, destiny, moral integration—onto horizontal substitutes—growth, power, consumption, tribal security—two parallel shifts tend to appear.

Externalities become normal—decisions are made as if only the short-term, measurable, local payoff matters. Long-range, diffuse consequences (environmental damage, health burdens, social decay, institutional brittleness, intergenerational debt, etc.) are pushed off onto third parties, underdeveloped countries, the public, or future generations.

Because these costs are externalized—socialized, postponed, or displaced—the machine can run smoothly for a time while quietly consuming the shared planetary environment it requires to endure. Some typical examples would include:

> Climate: greenhouse gases treated as free dumping into the atmosphere.
>
> Pollution/toxics: health and cleanup costs borne by communities and public budgets.
>
> Biodiversity/land use: ecological degradation treated as "outside" the ledger.
>
> Antibiotics: resistance created by overuse, paid later by patients and hospitals.
>
> Debt/short-termism: today's stability purchased by borrowing that constrains tomorrow.

Each vector in the list below is therefore presented in a two-part frame: result side / amplifier side.

The result side describes what a society's inner value-orientation produces on the outside: how a flattened or instrumental horizon shows up in everyday behavior, incentives, and material conditions.

The amplifier side describes the feedback: how those conditions, once in place (and any new shocks that hit them) then worsen the interregnum by intensifying tensions—spreading fear, shortening time horizons, hardening factional identity, and increasing the appeal of proceduralism, surveillance, and "salvation by force."

In the result–amplifier sequence, externalities belong primarily on the result side. They are what a horizontalized value-order produces when it treats the shared planetary environment as inconsequential and "off the books."

Then those externalities return on the amplifier side as accumulated harms—climate stress, health fragility, ecological volatility, fiscal brittleness—that *magnify* distrust, polarization, and coordination failure under interregnum conditions.

Meaning and narrative stressors

Result side: The culture becomes saturated with stories but starved of trust. Narratives multiply while suspicion hardens that every narrative is manipulation. Moral language collapses into preference and grievance, and irony replaces moral seriousness as the safest posture.

Amplifier side: The hunger for ultimacy returns in distorted "big stories"—conspiracy theories, cultic politics, and tech-salvation myths—that offer fast orientation. These counterfeits intensify fragmentation by turning shared reality into rival meaning-tribes.

Trust and legitimacy stressors

Result side: Institutions lose moral credibility, not merely public approval. People doubt that any authority can be rightful even when it is competent. Bad-faith attribution becomes the default, so disagreement loses the possibility of mutual intelligibility.

Amplifier side: The legitimacy vacuum produces authority whiplash. Rejection of authority appears alongside craving for strong arbiters when persuasion fails. Proceduralism rises as surrogate morality, deepening cynicism because process cannot replace purpose.

Personhood and relational stressors

Result side: The relational fabric thins: Loneliness increases as durable bonds weaken, producing high connectivity with low communion. Communities become transactional networks, and people speak through categories more than they address one another as persons.

Amplifier side: Relational precarity converts ordinary disagreements into identity threats that must be defended rather than questions that can be explored. The search for belonging becomes more vulnerable to tribal capture, accelerating conflict.

Freedom-pathology stressors

Result side: Freedom is redefined as non-accountability—autonomy without telos, choice without purpose, liberty detached from any shared good. The culture oscillates between permissiveness and control because restraint no longer feels intelligible from within.

Amplifier side: The oscillation produces responsibility fatigue: moral exhaustion, decision overload, and retreat from durable commitments. As commitment weakens, social trust thins further and coercive regulation becomes easier to justify.

Knowledge and epistemic stressors

Result side: Information expands while interpretation collapses. Data multiplies but synthesis is distrusted or unavailable, leaving facts without shared meaning. Truth becomes a weapon, and credibility attaches to tribe more than to evidence.

Amplifier side: Expertise is alternately discredited or idolized, producing either contempt for knowledge or submission to a secular priesthood. Time horizons shorten, planning yields to reaction, and collective action becomes fragile.

Moral and spiritual stressors

Result side: Ultimacy migrates from the vertical to the finite. Ideology, nation, identity, market, autonomy, or technology take on sacred weight. Religion, where it persists, is tempted into transaction and boundary, replacing interior transformation with technique or badge-wearing.

Amplifier side: Worship displacement attempts to make relativities behave like absolutes, so compromise feels like betrayal and opponents become existential threats. Moral life is outsourced to slogans, intensifying polarization and weakening self-examination.

Institutional and administrative stressors

Result side: Horizontal systems overreach. Institutions attempt to manufacture meaning they cannot generate by expanding management, messaging, metrics, and control. Policy churn accelerates—constant reforms with little legitimacy—so "solutions" fail to stick.

Amplifier side: Coordination failure becomes predictable: Even recognized problems resist collective action because trust is thin and motives are suspect. Dysfunction then justifies more management and enforcement, tightening the spiral.

Affective and aesthetic stressors

Result side: Ambient anxiety and anger become the background emotional weather, priming the nervous system for reaction. Attention fragments, stimulation becomes addictive, and the interior life is kept at bay by constant input.

Amplifier side: Chronic agitation makes societies easier to manipulate and harder to stabilize, so conflict escalates faster and settles slower. Beauty flattens into branding and culture into a feed, weakening depth and shared encounter with meaning.

Ecological Vectors as Amplifiers

Planetary heat-load and cascading impacts

Result side: A horizontalized value-order rewards throughput—consumption and energy use—while treating the atmosphere as a free sink whose costs can be ignored. The system externalizes long-horizon consequences because restraint lacks a shared moral gravity.

Amplifier side: Heat, drought/flood volatility, wildfire seasons, and coastal pressure multiply social strain through displacement, scarcity politics, and institutional overload. Stress becomes political accelerant, sharpening polarization and power contests because weakened integration cannot absorb the shock.

Biodiversity loss and ecosystem-service degradation

Result side: A flattened value horizon treats living systems as inventory to liquidate rather than a trust to steward, quietly removing the "free stabilizers" civilization relies on. Ecological functions—pollination, soils, water regulation, fisheries, disease buffering—are discounted until failure becomes visible.

Amplifier side: As those stabilizers degrade, volatility rises across food, water, and health, raising baseline anxiety and demand for control. A destabilized social commons makes substitute ultimates more magnetic, tightening the interregnum's drift toward tribal securities and quick certainties.

Water stress and food-system fragility

Result side: Short-term optimization overdrafts resilience—depleted aquifers, narrowed crop diversity, and mispriced water that treats a civilizational necessity as an endlessly substitutable input. Food systems become efficient on paper but brittle under heat and supply disruption.

Amplifier side: Scarcity turns policy into existential conflict: When communities doubt access to water, cooperation collapses into suspicion and governance shifts toward emergency logic. Polarization deepens and coercive control becomes plausible because shared conviction is too thin to stabilize restraint.

Climate-linked population displacement and migration pressure

Result side: Uneven development and extractive political economy concentrate risk, thin local resilience, and treat places and people as inputs to growth rather than communities to be sustained. As livelihoods erode, migration becomes structural rather than episodic.

Amplifier side: Mass movement presses on trust and shared narrative by forcing belonging, obligation, and identity to the front where moral vocabulary is already frayed. Under interregnum conditions, this strain converts easily into fragmentation and nationalist consolidation, seeking cohesion through boundary rather than meaning.

Mortgaged future and fiscal brittleness

Result side: Institutional short-termism borrows against the future to preserve present consumption, calm, or growth, turning governance into a narrowing corridor of tradeoffs. Debt service crowds out stabilizing investment and reduces adaptive capacity.

Amplifier side: When shocks arrive, constrained response undermines legitimacy. Publics infer capture or incompetence as governments oscillate between austerity and inflation risk. Fiscal stress then deepens social fracture, tightening the interregnum's stress-to-fear feedback loop.

Resource throughput escalation and pollution load

Result side: Materialism operationalizes growth as quasi-sacred, normalizing extraction while treating landscapes, oceans, and bodies as sinks for what the model will not price. Systems become tightly coupled as climate, biodiversity loss, and pollution reinforce one another.

Amplifier side: Pollution concentrates harm in socially legible ways, producing grievance ecologies of inequality, injury, and perceived abandonment. In interregnum conditions, these grievances escalate into distrust and factional conflict because harms are read as evidence of moral-order failure, not merely mismanagement.

Population health and fertility stressors from chronic disease and environmental exposures

Result side: Processed environments optimize for convenience and scale—food, chemicals, and built space that normalize sedentary lives—while discounting human flourishing and long-term vitality. The burden accumulates as chronic disease, mental strain, and care load quietly absorb resources and hope.

Amplifier side: Diminished health narrows the margin for adaptation: Households grow fragile, workforces less durable, and communities more dependent on strained institutions. That dependence accelerates legitimacy decay because systems asked to carry ever more load with thinning trust are experienced as failing arbiters.

In Summation

Result side: These vectors are the *material signature* of displaced ultimacy—what the shift looks like once it's engineered into production, consumption, and land use. When ultimacy is no longer anchored vertically, operating loyalties migrate into horizontal substitutes that maximize what can be measured, scaled, and competed for. Extraction becomes normal, externalities become invisible, and the social commons becomes nobody's responsibility in particular.

Amplifier side: The vectors then return as accelerants—heat, scarcity, disease, debt, and displacement—that thin the remaining tissues of trust, intensify factional fear, and deepen the authority vacuum at the level of meaning. The culture becomes both more likely to generate destabilizing pressures and less able to coordinate a coherent, morally serious response.

What emerges from the convergence of these vectors is not simply chaos but a brittle order—one increasingly tempted to manufacture coherence through surveillance, regulation, procedural overload, and emergency powers, because organic integration has weakened.

This is the phenomenology of the interregnum: a civilization still functioning on the horizontal axis, yet inwardly disoriented—trying to stabilize coordination without restoring ultimacy, and trying to enforce order without recovering the horizontal intelligibility that provides personal and institutional orientation. Recovery requires both: renewed vertical anchorage to ultimacy, and renewed horizontal orientation capable of guiding intelligent, compassionate action at planetary scale.

AI as a Meta-Accelerant

AI is not "another vector alongside the others." It is closer to a *meta-vector:* a general-purpose accelerator that couples to *all* the others we have listed.

AI relates to these environmental vectors in two simultaneous ways, and the difference between them is ultimately moral rather than technical. On one side, it functions as an optimization engine: It intensifies throughput, control, metricization, and acceleration, strengthening the very horizontal displacement already described by making extraction, persuasion, and management more efficient and more scalable.

On the other side, it can serve as an instrument of coordination and sense-making, improving foresight and reducing certain risks—but only insofar as it becomes governed by a recovered vertical axis, where ends precede means and restraint precedes capability. In that sense, AI does not decide our trajectory; it amplifies the one we have already chosen, either deepening the horizontal drift or becoming a tool placed under higher loyalties.

AI does not *cause* displaced ultimacy, but it increases its yield: It makes horizontal optimization so powerful that the secularized regime can prosper longer while continuing to hollow out the vertical axis that alone can command restraint.

The Interregnum-Specific Risk: AI as a Surrogate Vertical Axis

In assessing the interregnum's impact, the distinctive danger posed by AI is not merely technical; it is a migration of ultimacy. When the shared vertical horizon weakens, intelligence systems are easily sacralized—not as tools, but as substitutes for orientation itself.

They promise salvation by competence, the reassuring belief that the system will manage reality on our behalf; salvation by control, where surveillance and enforcement stand in for the slower work of moral consensus; salvation by explanation, where synthesis is mistaken for wisdom and prediction for purpose.

In each case, the appeal is not irrational. It arises from a genuine hunger for order and intelligibility—but it attaches that hunger to mechanisms incapable of bearing it. Authoritarian political regimes arise because desperate people will support anything that promises social order.

This dynamic reproduces the familiar pattern of proceduralism as surrogate morality, only now the procedure is automated and clothed in the aura of objectivity. Because the outputs are numerical, statistical, or computational, they appear neutral, inevitable, even authoritative. Yet the neutrality is partial and the authority borrowed, for the system optimizes what it is asked to optimize and cannot supply the ends it presupposes

The danger, therefore, is not that machines become powerful, but that a culture already uncertain of its moral center begins to treat their power as a final court of appeal.

AI as a Servant of Recovering Ultimacy

AI can also support recovery, but only if it is placed back into a restored hierarchy rather than allowed to become a rival horizon. The order is simple and severe: Ends first—truth, goodness, beauty, human dignity, long-horizon responsibility—and only then the means by which those ends are served: institutions, policies, and technologies, including AI. When that vertical ordering is intact, intelligence is disciplined by spiritual values; capability is constrained by conscience; efficiency answers to meaning.

In that posture, AI becomes what it properly is: a capacity tool. It can extend foresight, strengthen coordination, and improve stewardship across complex systems without pretending to be the source of purpose. It does not replace moral judgment with automation, nor does it compensate for legitimacy collapse with surveillance and control. It assists the work of a culture that understands its stewardship imperative and therefore can use power without worshiping it.

If ultimacy remains horizontalized, AI amplifies externalities, accelerates churn, and reinforces coercive substitutes for legitimacy—more optimization, more surveillance, more managerial control standing in for the slower work of shared conviction.

If ultimacy is re-centered vertically, the same capacities can extend foresight, coordination, and stewardship without being mistaken for the source of value. AI is not the new story: It is a powerful instrument that either serves a recovered narrative—or, in a vacuum of meaning, becomes a persuasive counterfeit of one.

Vertical Re-Orientation to Ultimacy

Interregnum recovery does not begin where late-modern instincts keep trying to begin: in institutional repair, media strategy, policy design, or new systems of coordination. Those are downstream effects. The interregnum is, at root, a crisis of misplaced ultimacy—the loss of a vertical horizon capable of ordering life.

When ultimacy is displaced, the culture does not merely become confused; it becomes value-anemic. People still speak constantly about "values," but in practice values are reduced to preferences, identities, procedures, or the moral theater of outrage. The result is not neutrality; it is drift. And drift always invites substitutes: Tribe becomes sacred, technique becomes salvation, coercion becomes legitimacy, consumption becomes consolation.

The corrective is reorientation to vertical ultimacy, because only that re-centers the interior life where spiritual values are actually generated. Spiritual values are not manufactured by institutions. They are not issued by committees. They are not produced by "culture" as an abstraction. Spiritual Values arise in indwelt *persons*—specifically, in persons who experience a reality higher than self-interest and higher than social reward—persons who then allow that reality to re-form what they love, what they fear, what they will sacrifice for, and what they will not do even under pressure.

This is why the decisive arena for the initial focus of action and resources is not the public square, but the interior one: conscience, worship, prayer, moral courage, mercy, truthfulness, and the slow renovation of desire. A person who has been vertically re-anchored becomes a living source of value: not merely someone with opinions, but someone whose life now carries moral gravity.

Once that is granted, the downstream sequence becomes intelligible. Institutions and culture conserve, amplify, and distribute the values already generated in persons. When the upstream source is healthy, institutions can function as instruments—schools that educate rather than indoctrinate, laws that embody justice rather than merely enforce power, economies that serve human flourishing rather than treat persons as inventory.

But when the upstream source is sick—when persons are inwardly unmoored—institutions cannot compensate. They may become more complex, more procedural, more surveillant, more punitive; yet they will still be hollow, because they are trying to stabilize life without a living value-source. In an interregnum, this hollowness becomes visible as legitimacy decay: People stop believing that institutions have moral authority, not merely that they are inefficient.

This is why "fixing society" by structural redesign alone is a category error. You can redesign the machine endlessly and still feed it the same fuel. Old motivations migrate into new forms. Corruption learns new vocabulary. Compliance replaces conviction. Symbolic victories outpace lived transformation.

In that condition, reform becomes a performance that cannot regenerate trust because it does not regenerate the moral substance that trust requires. Recovery cannot be engineered from above because the required material—shared value—cannot be fabricated at scale. It must be generated at its source, in persons, and then gradually normalized through relationships, communities, and practices that carry it forward.

Civilizational reorientation, then, is not first a campaign; it is *leavening.* It is the slow multiplication of transformed people living by transformed values, until the pattern of those lives becomes socially legible. When enough persons live from a vertically grounded value-order—truthfulness that costs, mercy that endures, courage without cruelty, humility without passivity—culture begins to regain coherence without coercion.

Shared moral vocabulary returns because it is being spoken in deeds, not merely argued in words. Institutions can again become expressions of a lived moral consensus rather than substitutes for one. In this sense, it is not civilization that makes the person; it is persons who make civilization—simply because persons are the originating site of spiritual values, and spiritual values are the only stable foundation on which any civilization can cohere

This is the hard, hopeful claim: The interregnum ends only when ultimacy is restored to the vertical axis, and the human sources of value are reactivated. Everything else—culture, institutions, the shape of public life—follows from that recovery, because it has always been downstream from it.

Horizontal Reorientation to Destiny

Vertical reorientation is indispensable, but it is not sufficient. The vertical axis restores the source of value—the inward anchoring by which conscience becomes more than social conditioning and worship becomes more than sentiment. Yet human beings cannot live by inward ultimacy in a vacuum.

If the interregnum is a crisis of displaced ultimacy, it is also a crisis of interpretive collapse, The shared stories by which persons once made sense of suffering, duty, hope, and progress have thinned or shattered. Under that pressure, even sincere spiritual renewal can remain fragile—private, episodic, easily overwhelmed by the roar of events—unless it is accompanied by a horizontal re-ordering.

An inhabitable universe frame spacious enough to interpret experience, organize meaning across time, and make the practices of love and service intelligible amid late-modern complexity is essential. Vertical ultimacy supplies the living fire; horizontal orientation supplies the architecture that holds that fire steady while life moves through history.

A coherent framework does more than explain; it stabilizes the moral life by giving it proportion. It names the world as a field of growth rather than a theater of cynicism. It grants suffering a context without romanticizing it. It renders duty humane, hope rational, and mercy brave. It empowers persons to love and serve in a troubled world not by offering escape, but by supplying a horizon in which faithful action remains meaningful when outcomes are delayed, obscured, or reversed.

In an interregnum, this is not decorative; it is structural. Without an inhabitable story-form, the regained vertical axis has difficulty finding traction in the horizontal world; it becomes harder to sustain courage without cruelty, truthfulness without panic, mercy without naïveté, and perseverance without illusion. The horizontal frame is the medium through which vertical life becomes durable, communicable, and socially legible.

The narrative in the Urantia Papers provides precisely this: a frame of reference in which values derived from vertical ultimacy can be applied with wisdom to the horizontal axis. It offers a worldview in which moral insight is not only privately consoling but publicly usable—capable of being translated into judgment, proportion, and responsibility.

This matters because we do not embrace spirituality in the abstract: We face decisions in history—how to respond to deteriorating environmental vectors and their cascades, how to weigh short-horizon relief against long-horizon stewardship, how to preserve human dignity amid displacement, scarcity, and institutional brittleness, and how to act without surrendering to panic, ideology, or resignation.

A workable worldview does not remove tragedy, but it prevents tragedy from becoming interpretive chaos. It gives moral agency a stable horizon in which to choose, endure, and serve. The Urantia Papers frame of reference strengthens this agency by adding intelligibility to responsibility. It places planetary stress inside a larger account of evolutionary progress, moral maturation, and the destiny of worlds, so that practical decisions can be made not merely as reactions to threat, but as contributions to the higher purposes of civilization's development.

But here the danger intensifies, because strong universe frames easily can become distorted into ideologies, can be converted into tribal identities. A narrative meant to function as a map can be turned into a flag. Instead of remaining an orientation that helps persons perceive, choose, and grow, the framework becomes a badge of belonging—an insider language, a status ladder, a boundary line separating "us" from "them."

When this happens, predictable dynamics follow: boundary maintenance (who is "really in"), status signaling (who has mastered the vocabulary), loyalty tests (agreement as proof of belonging), outgroup contrast (others cast as blind or inferior), and subtle recruitment pressure (growth measured by expansion rather than transformation). In that moment, the horizontal structure—meant to carry vertical ultimacy into life—quietly becomes a substitute-ultimacy of its own. Loyalty migrates from God and conscience to the group and its markers, and the interregnum's basic disorder reappears inside the very community that thought it had escaped it.

Escaping this trap is perhaps the greatest unrecognized challenge facing readers of the Urantia Papers today.

Implications for Dissemination

A mature dissemination strategy begins by distinguishing book *availability* from book *promotion.* Availability is stewardship: ensuring that a serious text can be found, accessed, afforded, and kept in circulation across languages and generations. Promotion, by contrast, easily slips into recruitment energy—persuasion, branding, identity capture, and the subtle implication that interest ought to culminate in belonging.

Under interregnum conditions, that latter posture is self-defeating: People are saturated with marketing and suspicious of agendas, so "book-first" pressure triggers defenses long before the content is encountered. The deeper we descend into the interregnum, the shriller become the cries of people who are hawking solutions. The correct center of gravity is therefore quiet access paired with patient formation: Make the book widely available, while refusing to treat salesmanship as an engine of spiritual transformation.

Moral and spiritual formation remains the highest priority because it preserves the rightful causal sequence: experience → values → orientation → durable inquiry → deeper study → enhanced service potentials. The Urantia Papers can function as a conceptual framework only when persons have some reopened capacity for the vertical axis—worship, moral seriousness, prayer, service, and interior transformation.

Spiritual formation is the non-threatening ecology in which curiosity can ripen into hunger and hunger into responsible engagement. It is sustained by hospitality, facilitation norms, mentorship, service rhythms, and "exit-easy" inquiry spaces where the first requirement is not agreement but freedom. In such an ecology the Urantia Papers are not wielded: They are discovered.

That is why "quiet infrastructure" is strategically coherent. We are not trying to *cause* questioning; we are trying to ensure that when questioning ripens into genuine seeking—often privately, episodically, and without social appetite—the materials of orientation are already present, non-coercive, and findable, so the person can take the next step without first having to step into a socio-religious container.

Distribution, then, should be organized as quiet infrastructure rather than aggressive market push. The aim is not to "convert the culture" through campaigns, but to ensure that when a person's inward questions mature, the text is actually there—close at hand, credible in its placement, and easy to obtain.

That means emphasizing translations, metadata and search visibility, stable supply channels, and long-term preservation. It means placing the book where serious seekers naturally go when they are already looking: libraries and library systems (public, academic, and special collections), retreat centers and spiritual renewal houses, chaplaincy and pastoral environments, university religion/philosophy shelves, prison and hospital libraries, interfaith centers, and select independent bookstores. The point is not that these venues "sell"—the point is that they legitimate encounter by embedding the text in ordinary pathways of inquiry.

Meanwhile, we invest our most precious energy not in marketing, but in spiritual formation: cultivating persons and small communities whose lives exhibit vertically transformed values, and who can host non-threatening inquiry without recruitment.

When distribution is centered on availability and formation is prioritized as the living engine, dissemination becomes both broader and cleaner, resulting in wider reach without distortion. Transformative assimilation can deepen without becoming entangled in social identity markers, badges of belonging, loyalty recruitment, in-groups/out-groups, and pressure to adopt a package rather than freely discover values.

Why Re-Anchoring Persons Must Precede Structural Reform

Re-anchoring does not mean fostering ideological uniformity. It means that enough individuals recover an ultimate reference, moral seriousness, narrative continuity, and responsibility beyond self-interest. Where these qualities begin to live again—even unevenly—institutions regain the capacity to *express* legitimacy rather than fabricate it: Policies and procedures draw authority from shared conviction instead of enforcement or symbolism.

Without this re-anchoring, structural repair produces compliance without commitment, efficiency without meaning, and order without legitimacy, as institutions become substitutes for value rather than containers of it.

In short: Institutions can distribute values widely, but only persons can generate them. Civilizational progress is therefore not primarily a legislative or administrative achievement. It is the cumulative externalization of innumerable interior transformations.

Enlistment and Membership Are Counter-Productive

They shift the center of gravity

What begins as interior value formation—conscience clarified by experience, worship deepening into service—gets pulled outward into the machinery of affiliation. Attention migrates from the slow work of personal transformation to the faster work of boundary maintenance: Who is *in,* who is *out,* what language marks belonging, what gestures prove loyalty.

The person's spiritual labor is subtly replaced by social positioning, and the framework is no longer primarily a means of orientation; it becomes a badge. In that shift, the very thing we are trying to cultivate—free, inward, morally generative life—gets displaced by identity management and the signaling economy that inevitably attends membership.

They change the question

Instead of asking, "Is this true—does it illuminate reality, clarify conscience, and integrate my experience?"—the seeker is nudged toward a different, socially loaded inquiry: "Do I belong with these people?"

That second question does not primarily engage the moral and intellectual faculties. It triggers social psychology—status, similarity, reputational risk, fear of exclusion, desire for approval. The framework is no longer evaluated as orientation architecture but as a social proposition, and once that shift occurs, assent becomes entangled with identity protection. What should have been philosophical and spiritual discernment is recoded as a belonging decision, and the path of inward transformation is replaced by the calculus of affiliation.

They trigger defensive heuristics in observers

To outsiders, recruitment and membership cues are rarely interpreted as neutral: They are read as bids for power, identity expansion, or social capture. In a culture conditioned by countless ideological and sectarian appeals, the presence of enlistment mechanisms activates pre-rational defenses: *What do they want from me? What will it cost? What am I being asked to surrender?*

At that point the content of the book becomes secondary to the behavior of the group around it. The public does not evaluate a framework in the abstract; it evaluates the social ecology it generates. When the ecology signals boundary pressure and loyalty, the material is filed away under "movement," and its civilizational usefulness is dismissed before it is ever tried.

They create maintenance demands

Once a group is structured around membership, it must continually manufacture cohesion: Leadership roles solidify, conflict requires adjudication, norms have to be enforced, and the shared narrative must be defended against internal drift and external critique.

A great deal of psychic and organizational energy is then diverted into preserving the collective as a collective—keeping people aligned, keeping dissent manageable, keeping the boundary intelligible.

What began as the articulation of meaning becomes the administration of belonging. The framework's purpose subtly inverts: Instead of serving persons as an aid to orientation, persons are recruited to serve the group as the custodian of the framework.

The more precise principle is not "avoid all association," but rather: Do not make affiliation the primary pathway to engagement. There is a real difference between community as the by-product of shared inquiry and community as the prerequisite for access or validation.

The danger is not in friendship, testimony, or joy themselves, but in the subtle change of function: what begins as communion becomes enclosure, what begins as witness becomes enlistment, what begins as ministry becomes administration.

When gatherings remain permeable, they function as reading circles or philosophy salons rather than movements—spaces where ideas can be tested, savored, and carried away without any implied transfer of allegiance

The challenge is to preserve a culture of interiority and openness so that when enthusiasm and fellowship naturally arise, they express as hospitality and conversation rather than enlistment and boundary-formation.

A Urantia Papers Strategy for the Interregnum

The Urantia Papers do not *create* the values they illuminate; many are native to personal repercussions of Divine indwelling—present across religious history. The revelation's distinctive role is to *clarify and intensify* those values in continuity with the teachings of Jesus and to *re-secure their intelligibility* under late-modern conditions of fragmented knowledge.

Framing the problem correctly: avoiding the cultic cocoon

The Urantia Papers constitute a universe frame meant to be inhabited. They provide an instrument of orientation to be taken up the way one takes up a compass, a grammar, or an architecture—something that clarifies experience, steadies conscience, deepens worship, and renders love and service intelligible amid the world's pressures.

A cultic cocoon, by contrast, is what forms when the same material is handled as identity rather than use. Then the social ecology begins to signal enclosure—insider language, boundary vigilance, interpretive gatekeepers, quiet pressures to adopt a package—and modern observers read the most available category for such signals as *cultic.*

The decisive difference is not the size of the cosmology but the center of gravity of the reader community. In framework mode, the emphasis is experience, discernment, translation into public grammar, and leavening through personal living; in cocoon mode, the emphasis slides toward recruitment logic, status markers, and loyalty maintenance.

A universe frame remains non-cocooning when it refuses to solve the human hunger for belonging by manufacturing an in-group, and instead stays faithful to its proper work: to be used—freely, humbly, and fruitfully—as a spacious interpretive house in which persons can mature before God without having to "join" a new social world to do so. The question is how to re-situate the text so its architecture can do its work in persons and communities without forcing an identity threshold that the modern world correctly mistrusts.

A critical constraint from the Urantia Papers

The Urantia Papers are unambiguous on two strategic constraints that cannot be negotiated away. First, institutions are not the primary agents of transformation. They are downstream from value's true source in personal experience and are readily bent out of shape in an interregnum.

Secondly, the religion of Jesus advances by lived experience, quiet leavening, and personal embodiment.

From this it follows with near-mechanical force that any effort to "promote" the book as a movement, an identity, or a cause will only deepen its marginalization and thicken the cultic aura around it. That is not a messaging defect; it is a category error—treating an orientation architecture as though it were an organization to recruit for.

The real task, then, is more exacting and more luminous: to reposition the papers without promoting the book, so they can work as they were meant to work—through persons, in life, as a steady grammar of meaning that spreads not by pressure but by usefulness.

The question is not "How do we get people to read the Urantia Papers?" The question is: How do we allow its worldview to become culturally available without requiring prior engagement with readership cults or Urantia institutions?

Decenter the book as an object; foreground the framework

The Urantia Papers will not spawn a new civilization by winning the contest of assent—as though history will turn on how many people own copies of the book or "believe." It will do its deeper work by becoming quietly useful: an *architectural orientation* rather than a doctrinal system, a *personalist cosmology* rather than a competing religion, a framework for *inhabiting modernity* rather than an alternative identity

In that register it stops being marketed as "the answer" and begins to function as a disciplined way of seeing—of situating the self, integrating experience, and restoring proportion on the vertical axis amid horizontal turbulence

This is why the emphasis on orientation and an inhabitable framework is the decisive hinge. The book becomes transformative precisely when it is treated less as a badge to wear than as a world to enter, a set of load-bearing coordinates by which persons can generate steadier values, clearer meanings, and more humane forms of common life.

Lead with problems modern people already feel, not with revelation claims

People living in an interregnum already carry the symptoms in their own nerves and speech: teleological collapse, identity exhaustion, institutional mistrust, future-anxiety, and the ongoing fracture between science, meaning, and moral life.

The Urantia Papers speak directly to these conditions. The entry point is not, "Here is a revelation you need to read," but, "Here is a coherent way to live in a personal universe after secularization, after science, after institutional failure—without rejecting what is true in any of them: science's explanatory power, secularization's exposure of nominal religion, and institutional failure's demand for discernment—while refusing to let any of them become the final horizon of meaning."

Let the book arrive as a resolution to lived problems, a restoration of proportion and integration where modern consciousness has been disjointed, rather than as a bare truth claim that requires assent or engagement with socio-religious complexities before it can yield fruit.

Replace "Urantia identity" with Urantia competence

The cultic cocoon persists for a simple reason: Urantia institutions are too often framed as something to *belong to*—an identity to adopt, an institution to support. The exit appears when the revelation is framed as something to *think with:* an orientation architecture that can be tested in the light of lived experience.

The decisive reader's question should not be, "Do I believe in the Urantia Papers?"—as though the primary act were allegiance—but rather, "Does this framework actually help me understand reality, myself, and the future better than the alternatives?" That is a competence test, not a loyalty test. It invites discernment instead of enlistment. It allows the book's value to be discovered as intelligibility, proportion, and integration rather than a portal to group affiliation.

The decisive role of mediating works

Historically, major religious insights do not step out of their cocoons full blown by a direct appeal to the public. They exit gradually by way of secondary works—books, music, sermons, theatrical productions, personal conversations—that *translate* and *contextualize,* that de-escalate the sense of totality, and that make the material habitable for outsiders who did not ask to cross an identity threshold.

The hard but necessary conclusion follows: If the Urantia Papers are promoted directly, they will remain marginal. If they are lived, translated, and architecturally mediated through secondary works relative to people's concerns, they will slowly shed their cultic profile as the aura of "movement" is dissolved by the usefulness of the orientation they provide.

The evangelical imperative in the Urantia Papers is fundamentally vertical: Regenerate persons, recover the living Jesus in experience, and let that reality become socially consequential downstream—while refusing the horizontal shortcuts that mimic renewal through promotion, branding, recruitment, or institution building.

The irony is that the revelation will aid the birth of a new civilization only when it stops trying to be noticed as the midwife.

The Social Genesis Sequence

Because Urantia Papers readership as a socio-religious movement is historically young, a strategic danger is to confuse visibility with maturation, and to attempt creating at the horizontal level what can only be generated upstream.

The Social Genesis Sequence states a simple priority: Spiritual values are born only in persons—through interior transformation, stable vertical orientation, and the slow work of inhabitation in which the tacit self-story is re-authored and character becomes reliable under ordinary pressures.

Everything downstream is secondary: trust-mediated diffusion through intimate relationships and networks; the gradual thickening of shared practices in service, worship, stillness, and truthful dialogue; and only later the codification of those practices into institutional routines and the emergence of public story.

For that reason, an early "campaign" aimed at institutions, public legitimacy, or civilization-scale influence is not merely premature; it is structurally misordered and purpose defeating. When one tries to start at sequence stages 6–8 without mature substrates in sequence stages 1–3, the typical result is brittle identity, boundary-making, and persuasive rhetoric substituting for formation—an imitation of renewal rather than renewal itself.

Our aim, therefore, should not be to launch a movement, but to foster persons and small relational ecologies in which the book can be quietly internalized as orientation architecture. If wider social effects ever come, they will come as downstream fruit, developing organically rather than as a managed program. Focus on the upstream task and let downward diffusion follow evolutionary processes.

The Social Genesis Sequence is the path by which spiritual values migrate from their origin in persons, through cultural diffusions, and finally to their crystallization in institutions. We can imagine this sequence as describing the progressive amelioration of interregnum conditions.

This is the sequence of spiritual value migration in a culture:

1. Spiritual values originate in personal spiritual experience
2. Spiritual values generate meanings in proto-narratives
3. Spiritual values diffuse into family and intimate relationships
4. Spiritual values diffuse through social and business relationships
5. Spiritual values start to become incorporated into community practices
6. Community values become institutionalized in civic associations
7. Cultural integration is supported by values in shared narratives
8. Global civilization vertically oriented and horizontally progressing in all domains

Below we review this sequence in more detail, showing interregnum-related symptoms and Urantia Papers leavening.

1. Spiritual values originate in personal spiritual experience

What it is: The irreducible origin-point. Spiritual values are generated only in lived personal experience—in conscience, worship, moral choice, and the willingness to do the will of God as one understands it.

Strategic implication: Do not begin with campaigns, institutions, or "impact." Begin with formation, soulcraft: interior reorientation that yields reliable character and a stable vertical axis.

Failure mode under interregnum conditions: The person is de-centered by fear, status, ideology, distraction, or procedural busyness; value is outsourced to tribes or systems.

Urantia Papers leavening: The discovery of values is the basis for self-generated narrative architecture that deepens God-consciousness, enlarges destiny-horizon, and stabilizes moral will.

2. Spiritual values generate meanings in proto-narratives

What it is: The largely unspoken story the person lives inside: what is real, what matters, what counts as success, what suffering means, what kind of agent "I" am.

Functional role: Proto-narratives govern value-recognition and value-choice by preloading perception, motive, and the range of "plausible" actions.

Strategic implication: Renewal requires inhabitation and slow settling: re-authoring the tacit story so that vertical ultimacy becomes experiential center rather than an idea.

Failure mode under interregnum conditions: Ambient cultural scripts (cynicism, ideology, consumer status, fear media) colonize the tacit story and displace ultimacy.

Urantia Papers leavening: Spiritual values function as interior narrative architecture for re-authoring proto-narratives—restoring proportion, hope, and a coherent personalist horizon.

3. Spiritual values diffuse into family and intimate relationships

What it is: The first real proving ground. Intimacy tests whether transformation is durable: patience, truthfulness, restraint, forgiveness, reliability.

Functional role: Trust is the primary transmission medium; imitation precedes instruction; household life becomes the micro-ecology where values become enacted.

Strategic implication: If formation cannot stabilize here, it will not scale; domestic credibility precedes public credibility.

Failure mode under interregnum conditions: Fractured attention, economic stress, and ideological polarization thin trust and make intimacy brittle or controlling.

Urantia Papers leavening: Values expressed here as lived mercy and steadiness—not as doctrinal policing or identity branding.

4. Spiritual Values diffuse through social and business relationships

What it is: The wider field of meaningful ties beyond intimacy—friendship circles, colleagues, neighbors, mentors, recurring "third places."

Functional role: Reinforces values through repeated contact; carries tacit wisdom by proximity; bridges social worlds; enables "benign viral" diffusion without campaigning.

Strategic implication: The primary vehicle of spread is relational credibility, not messaging. Values travel fastest along trust-lines.

Failure mode under interregnum conditions: Networks become thin, performative, and tribal; outrage and identity signaling displace truthfulness and patience.

Urantia Papers leavening: Spiritual Value most often enter second-hand through the non-anxious presence of persons who have internalized their horizons.

5. Spiritual Values start to become incorporated into community practices

What it is: Repeated shared actions that train persons and pattern relationships—values embodied in time, not merely discussed.

Functional role: Builds trust, calibrates conscience, produces norm-emergence, and forms habits of love, duty, courage, and restraint.

Strategic implication: Practices are the bridge between private sincerity and durable culture. Study groups belong here as practice-concentrators—stabilizing shared language, norms, and mutual encouragement.

Failure mode under interregnum conditions: Practices degrade into events, activism without interior anchoring, or talk without obedience to truth.

Urantia Papers leavening: Supports value diffusion by reinforcing a service ethic.

6. Community values become institutionalized in civic associations

What it is: The stabilization of what works into structures that outlast founders: schedules, training pathways, procedures, standards, leadership succession, organized care.

Functional role: Preserves and replicates formation, prevents drift, increases care capacity, and provides continuity across generations.

Strategic implication: Institutions are *downstream fruit,* not the engine. Build them only as practices mature; otherwise you institutionalize immaturity and create brittle identity-systems.

Failure mode under interregnum conditions: "Legality without legitimacy"—procedure substitutes for conscience; incentives select careerism; survival replaces telos.

Urantia Papers leavening: Values influential in upstream feedback, shaping leaders' conscience and priorities.

7. Cultural integration is supported by values in shared narratives

What it is: The society-wide meaning layer—publicly available stories and symbols that teach what is real, what matters, and what counts as human dignity.

Functional role: Legitimation of authority and sacrifice; pedagogy of moral perception; symbolic compression; continuity through public memory.

Strategic implication: Public story cannot be manufactured credibly ahead of lived moral reality; it must arise from a stabilized practice-and-character base or it becomes propaganda.

Failure mode under interregnum conditions: Fragmented story fields compete for ultimacy; cynicism and manipulative rhetoric proliferate; moral incoherence deepens.

Urantia Papers leavening: Indirect reservoir underwriting public story via exemplars and secondary works—never as a membership test.

8. Global civilization vertically oriented and horizontally progressing in all domains

What it is: The transnational layer where interdependence binds societies—finance, media, migration, science/tech, ecology, and security architectures.

Functional role: Coordination for problems no nation can solve alone; the highest-order arena where legitimacy, restraint, and truth norms either converge or collapse.

Strategic implication: Civilization-level coherence is the *furthest downstream* effect. Trying to "go global" before upstream maturation produces brand-making, factional rivalry, and brittle institutionalism.

Failure mode under interregnum: Connectivity without shared moral integration—high interdependence, low trust; procedural coordination without legitimacy; competing civilizational stories.

Urantia Papers leavening: Contributes through narrative which illuminates a destiny horizon and personalist values which reinforce cross-cultural unity. It still must be understood as an orientation architecture rather than a cultic identifier.

The Social Genesis Sequence names a structural law of renewal: You cannot manufacture downstream maturity by amplifying upstream absence. Because Urantia Papers readership as a socio-religious movement is historically young, its standing danger is to confuse visibility with maturation—to attempt at the horizontal level what can only be generated *upstream* in persons.

Spiritual values are born only in personal spiritual experience—in conscience, worship, moral choice, and the lived willingness to do the will of God as one understands it—and they reshape the proto-narratives a person lives inside, re-centering life on vertical ultimacy until character becomes reliable under ordinary pressures.

Everything else is downstream and secondary: Spiritual values migrate along trust-lines through intimacy and relational networks, thicken into shared practices (service, stillness, truthful dialogue), and only later crystallize into institutions and public story

Any early "campaign" aimed at civic legitimacy, institutional footprint, or civilization-scale influence is structurally misordered: It tends to substitute rhetoric, boundary-making, and brittle identity for formation—an imitation of renewal rather than renewal itself. The strategic aim, therefore, is not to launch a movement but to foster persons and small relational ecologies in which the book can be quietly internalized as orientation architecture, letting diffusion and wider effects emerge as organic fruit rather than managed program.

A brief case study illustrates the role of upstream formation when integrative public order fractures: the post-Roman West and the monastic stabilization of moral and cultural continuity. This is illustrative rather than exhaustive. It is not a claim that monasticism "caused" medieval Europe; it is a focused illustration of how upstream formation can preserve and transmit values when macro-institutions fracture.

Case Study: The Post-Roman Interregnum and the Monastic Generation of Spiritual Values

The interregnum which emerged in the late-imperial and post-imperial West (c. 235–600 CE, with long aftershocks into the 7th century) has been extensively studied.

The collapse was not merely administrative. It was felt at all levels of civilization. The Roman system had supplied a shared coordination grid—law, taxation, roads, offices, civic identity, and a recognizable "public order." As that grid thinned and then fractured, society did not simply become neutral. It reloaded ultimacy into nearer substitutes: kin and patronage, warlord security, ethnos, local cult, coercive enclave, and improvised legitimacy. The result was not simple chaos but a contested field of rival stabilizers—many of them effective in the short run, none capable of sustaining a humane moral horizon at scale.

The monastic movement matters here as illustrative of the Social Genesis Sequence. Spiritual values originate in persons before they can be conserved by institutions and shape social expressions.

Early Western monasticism functioned as a disciplined ecology of vertical reorientation—conversion of will, purification of desire, prayer, obedience, stability, and service—organized not as a theory but as a repeatable way of life.

In an interregnum, such practices generate a rare social product: *reliable persons.* Where coercive order is intermittent and legitimacy thin, the most valuable "infrastructure" is not first a law code but the steady presence of people whose conscience has been trained to hold under pressure—truthfulness, restraint, mercy, patience, and duty sustained without external enforcement.

From there, the diffusion sequence becomes visible. Values stabilized in persons, thicken into communities of practice—monasteries ordered by rule, time, and shared labor—where trust can accumulate and be transmitted. These communities then extend outward along relational networks through hospitality, teaching, care for the vulnerable, mediation, and the slow normalization of moral life.

Monasticism can be read as one of Europe's great restorative experiments. A disciplined retrieval of the vertical axis in an age when the public order had fractured and the old civic integrator no longer held. It began as a school of inward spiritual allegiance—conversion of will, purification of desire, prayer, restraint, mercy, work—embodied in persons and stabilized in a community of practice.

In that early register it functioned as quiet infrastructure for a bruised civilization: producing reliable people, conserving memory, extending hospitality, and reintroducing moral continuity where coercive order had become the default substitute for legitimacy.

But the very success of such a formation ecology carries a built-in hazard. Once it scales into the institutional milieu—once it acquires property, jurisdiction, financial resources, political influence, status, and a civilizational role—it is pulled into the horizontal demands of administration: governance, boundary maintenance, enforcement, reputation management, political alliance, and the endless triage of complex social systems.

Here the decisive warning emerges: Any successful formation ecology, once scaled into institution, must actively defend its upstream spiritual center—or it will drift into management, boundary, and control. The drift is not first a moral collapse; it is a gravitational slide. Horizontal burdens multiply under pressures of scale, and if the vertical orientation is not deliberately protected and maintained, the institution slowly begins to live from what it manages rather than from what it worships—substituting procedure for conscience, unity for truthfulness, and organizational survival for the slow work of interior transformation.

The lesson is not cynicism about institutions—some institutional form is always necessary—but sobriety about their tendency to become ends in themselves. If the vertical center is not thoughtfully maintained, the horizontal will always expand to fill the field, draining the very inward life the institution was created to protect.

Strategy for Supporting Planetary Healing

If the objective is re-orientation to vertical ultimacy, then distributing the Urantia Papers is only one tactic within a broader strategy of cultivating environments which catalyze personal spiritual value-generation. Book publication, translations, and distribution remain essential because an architecture cannot shelter anyone who never finds it. Yet distribution is never sufficient, because what ultimately matters is whether the framework is actually *used:* internalized as a way of seeing that generates living values rather than merely multiplying copies.

What follows are suggestions for a strategy by which revelation ambassadors might participate in one of our planet's most amazing and enthralling epochs of social readjustment, moral quickening, and spiritual enlightenment.

Support convergent processes

Assume that the purpose of the revelation is to guide humanity through the interregnum and the way to accomplish that is to reorient humanity to the necessity for vertical calibration to ultimacy. Any activity that deepens conscience and self-examination, fosters reciprocal community, cultivates worship or contemplative interiority, strengthens responsibility expressed in service, and restores the integration of reason with meaning—should therefore be logically supported.

These practices may appear under explicitly religious, philosophical, or civic forms, but the affiliation is irrelevant. The criterion relate to *orientation to ultimacy,* the steady re-centering of the interior life toward what is highest and most real.

Interfaith outreach, therefore, cannot remain at the level of agreeable pluralist sentiment—an "I'm-ok-you're-ok" détente that flatters everyone while transforming no one. It should function as a convergent reinforcement of lived spiritual experience: the encouragement of inward depth, moral seriousness, and the vertical integration by which persons become reliable generators of spiritual values, capable of stabilizing community without coercion.

Far more fruitful than interfaith outreach might be *ecumenical* outreach, particularly in light of the Vatican II Decree on Ecumenism—the mandate to foster unity across the entire family of Christian believers regardless of doctrinal and liturgical differences.

Finance environments, not outcomes

Funding should be aimed at *conditions* rather than *conclusions:* creating spaces for study, dialogue, silence, and service; supporting translation and accessibility so the material can be encountered without gatekeeping; financing secondary works, placement of the book and its derivatives in places where it can be engaged without social pressure.

This posture refuses to purchase outcomes or manufacture allegiance. It underwrites the environments in which conscience can deepen, reciprocity can thicken, and living values can arise on their own terms—non-coercive, durable, and immune to the distortions of programmatic engineering.

A distributed social ecology

If the objective is vertical calibration rather than movement-building, then group identity must be kept structurally minimal—and the cleanest way to do that is to diversify the social infrastructure instead of consolidating loyalty to centralized institutions.

A diffuse ecology of small groups, public venues, service projects, study circles, interfaith dialogues, and contemplative practices prevents any single hub from becoming the gatekeeper of meaning, status, or access. Variety interrupts the formation of insider hierarchies and reduces the pressure to signal belonging; minimizes the rise of liturgical excess, lets persons participate without crossing identity boundaries, and facilitates keeping the framework oriented toward use rather than allegiance.

Historically, the diffusion of small renewal groups embracing revelation—house gatherings, circles of prayer, disciplined fellowships, quiet communities of practice—has helped essential spiritual teachings to survive centuries of hostility to invasive new systems of belief and practice, sometimes from political regimes and sometimes from religious institutions themselves.

What cannot be easily centralized cannot be easily seized; what lives in persons and in small communities cannot be eradicated by capturing a headquarters, silencing a pulpit, or co-opting a hierarchy. The result is resilience: a form of spiritual life that endures under pressure because it is dispersed, non-exclusive, and anchored in personal moral agency rather than institutional direction.

History also supports a decentralized, diffused ecology as a means of eliminating destructive power accumulation and financial abuse on the part of centralized institutions. The failure of sufficient diffusion has already generated persecution of readers in underdeveloped parts of the world—and the interregnum is just getting underway.

It would seem prudent at the very least to exercise caution and forethought in any thought about creating high-profile infrastructure at this early stage of the interregnum. Think in terms of being like those little mammaliaforms scurrying around the feet of dinosaurs which were heading for extinction.

Plural reinforcement preserves personalism

If value is discovered personally, then multiple channels and networks of development are not dilution. They provide valuable redundancy in service of freedom—many doors into the same interior work, so no single doorway becomes a gate or the target of assault by hostile forces.

Plural reinforcement preserves personalism because personalism rises or falls on a single principle: Value must be inwardly discovered and freely owned, and transmitted through interpersonal relationships—not socially installed. When a framework is socially diversified the result is many points of entry into the same interior architecture—no single channel can claim, "This is the way in," and no authority can convert access into political or financial leverage.

That redundancy is not dilution; it is protection: It prevents the framework from becoming a gateway. It prevents the rise of "sacerdotal sycophants," status ladders, or control systems. The moment any one pathway begins to harden into fundamentalism, technique, hierarchy, or identity pressure, other pathways can continue to offer discernment, conscience, and lived calibration.

In this way plurality keeps the emphasis where personalism requires it; not on allegiance to an enclave, but on the growth of persons who generate values from within and carry those values outward in enlightened service.

Resilience is increased because meaning is carried by many forms, not one fragile pipeline; and it prevents any single narrative from hardening into technique or control, where the framework becomes leverage to manage people instead of an architecture that helps persons generate living values for themselves.

Measure health indirectly

Avoid metrics that belong to recruitment—"conversions," "adoptions," "numerical capture," number of study groups, numbers of Society members—because they reward the wrong kind of success and quietly reintroduce the logic of movement and control.

Look instead for the slower, more legible symptoms of strengthened orientation: thicker local relationships that persist beyond enthusiasm; sustained volunteerism and service that outlasts mood; a renewed moral vocabulary capable of naming obligation and goodness without embarrassment; increased cross-group dialogue that does not collapse into sentimentality, and durable communities of practice that hold attention over time.

These are the signs that something real is happening beneath the surface—that the value-generation loop is active, that interior life is thickening into shared life, and that culture is beginning to regain its lost bearings.

Support the arts and keep them independent

Religious movements rarely spread by argument alone; they spread when a way of life becomes *imaginable,* and the arts are one of the primary instruments by which imagination is formed. Art gives a movement a sensory body: melody, image, gesture, architecture, story, and ritual texture that can be entered before it is fully understood.

It lowers the threshold of access by letting people *feel* a world—its moral atmosphere, its hope, its nobility—without first mastering its doctrines. In that sense, the arts translate vertical orientation into publicly shareable form. They make devotion legible, make meanings memorable, and create "thick" experiences in which conscience and desire can be re-patterned over time.

The arts also function as carriers across social boundaries and generations. Songs travel where books cannot; images can cross literacy barriers; drama and poetry can smuggle complex convictions past the defenses that reject overt persuasion. Because artistic forms are repeatable—sung, performed, reenacted, displayed—they become cultural memory systems, preserving a movement's emotional tone, exemplars, and moral grammar long after the founding intensity fades.

When this works well, the arts don't replace spiritual experience; they *tutor attention* toward it, giving the heart a vocabulary of reverence, repentance, courage, mercy, and joy, and giving communities a shared rhythm of recognition.

But the same power carries an interregnum-specific danger: Art can become a surrogate vertical axis. When aesthetic intensity is mistaken for spiritual depth, the movement drifts into spectacle—identity theater, branding, and emotional management—because beauty is easier to scale than holiness, and performance is easier to measure than character.

The cautionary rule is simple: When the arts are governed by the vertical center, they serve formation—deepening worship, enlarging moral seriousness, and strengthening humane imagination. When they are governed by horizontal incentives, they become propaganda, recruitment technology, or entertainment, and they quietly drain the very interior life they were meant to express.

Patience, faith, prayer, and trust

Don't lose sight of the fact that the Urantia Papers project dissipation of interregnum conditions as a thousand year undertaking.

A Fifth Epochal Dissemination Ethic

Pray, collaborate, act.

The cultivation of an active prayer life and a relationship with one's spiritual benefactors are crucial to wise dissemination. Vast spiritual resources are available to assist us in our task. We are not alone—neither personally nor collectively. Strategy may vary significantly relative to local dissemination environments.

Honor the primacy of the Urantia Papers.

Derive strategy from careful study of the revelation itself combined with wise analysis of the cultural milieu into which it needs to be diffused. Eschew apocrypha.

Honor the primacy of the person.

Treat living values as arising in persons through conscience, reflection, worship, service, and responsible freedom. Do not presume that any institution, text, or program can generate values in place of personal experience. Let outreach invite orientation rather than compel adherence.

Choose orientation over recruitment.

Do not aim at expanding membership, ideology, or brand. Aim instead at strengthening each person's connection to ultimacy—the credible vertical axis by which meaning, truth, and responsibility become livable. Refuse to treat numbers as the measure of success; look for depth of lived value.

Practice invitation, never coercion.

Keep outreach voluntary, transparent, and respectful of freedom of conscience. Reject pressure, manipulation, and fear as methods. Make the means embody the end.

Reinforce plural channels of formation.

Support the cultivation of vertical orientation through convergent practices—religious devotion, philosophical inquiry, community service, and artistic expression. Strengthen any process that deepens interiority, reciprocity, and responsibility, regardless of institutional affiliation without promoting preconceived programs.

Build an ecology, not a campaign.

Prioritize long-horizon cultural conditions in which value-generation can persist: Create access to reflective spaces, provide opportunities for dialogue, support study and translation, and sustain environments that encourage moral seriousness and mutual respect. Cultivate an ecology of orientation rather than a surge of attention.

Keep the scope humble.

Do not imagine that any single narrative, organization, or generation completes the work of orientation. Participate as a steward in an ongoing civilizational process of rediscovery and renewal. Prefer stewardship to possession.

Preserve integrity between means and ends.

Maintain coherence between message and conduct. Approach truth with intellectual honesty, goodness with service, beauty with care, and freedom with responsibility. Treat contradiction between message and conduct as reputational failure and spiritual error.

Track indirect measures of health.

Do not look for ideological conformity as evidence of success. Look for strengthened relationships, renewed moral vocabulary, sustained service, increased trust, and communities capable of self-correction.

Align with planetary rehabilitation

We are not in this alone, and the fifth epochal revelation is not confined to a bound volume or a human program. The book is one instrument of disclosure, but the larger movement of planetary rehabilitation is administered through living agencies: the Spirit of Truth, the ceaseless ministries of seraphim, the stabilizing services of the midwayers, and a superhuman planetary government tasked with planetary rehabilitation.

In the above context, our task is neither to "make something happen" by social force nor to engineer outcomes by technique; it is to cooperate intelligently and faithfully with an already-operative plan by becoming usable persons on the ground—steady in worship, clear in conscience, practical in service, and unthreatened by slow time.

We should therefore pray less for dominance and more for guidance: for discernment of where we are needed, restraint where we are not, and the humility to act as foot soldiers of light who assist higher ministries without trying to replace them, trusting that the long arc is carried by more than human hands.

We do not need to predict what *will* happen in order to establish a viable strategy. We are here attempting to identify what remains true regardless of *what* happens.

A viable plan must be flexible enough to change with circumstances; an invariant strategy does not. Institutional form must follow lived values as they emerge in the readership and their cultural contexts. We cannot forecast social outcomes, nor a can we manage history. Our challenge is to remaining rightly oriented under any outcome—preserving coherence precisely when conditions are least coherent.

The Decisive Focus: Soulcraft

We use Catholic "formation" in what follows as a reference model for a simple reason: It is the most fully developed, institutionally articulated account in Christianity of how a person is shaped over time into mature spiritual life and stable character. It is not chiefly "learning more religion," and neither is it the acquisition of ministry skills. It is the cultivation of a whole person—mind, heart, habits, relationships, and spiritual orientation—so that identity becomes stable, coherent, and fruitful over time with a consistent alignment to vertical ultimacy.

In Urantia Papers terms, formation—by definition—is orientation to the leading of the Divine Indweller, the Thought Adjuster: The steady education of conscience, desire, and choice under an interior divine presence whose aim is to align the human will with the will of God.

Read through the lens of interregnum recovery, formation is the upstream mechanism by which the Social Genesis Sequence can even begin its task of value diffusion. In Urantia Paper terms, that reconstitution is the slow establishment of a practical hierarchy: God's will is accepted as the organizing center, and the Adjuster becomes the primary interior reference point for value-recognition, moral seriousness, and worshipful loyalty.

In evangelical and broader Protestant usage, *discipleship* aims to form a person into a practicing follower of Jesus: A life in which beliefs, habits, character, and decisions are progressively aligned with Scripture and the perceived lordship of Christ. It strives for durability rather than mere enthusiasm—steady obedience, moral coherence, relational maturity, and a lived capacity to love God and neighbor in ordinary responsibilities.

In many settings it also includes "mission" as a normal outflow: not only personal holiness but a readiness to serve, to witness, and to participate in the community's work in the world. In our usage here, discipleship is one variant expression of the same interior reality: an apprenticing of the whole self to the will of God, increasingly mediated through the inward guidance of the Adjuster rather than through external pressure or identity capture.

Formation, discipleship, and soulcraft in a time of interregnum constitute the upstream work of cultivating persons and practices so that vertical ultimacy becomes experientially real and values are regenerated from within. From there spiritual values can be diffused into the larger culture along the lines of the Social Genesis Sequence—ultimately leading to the transformation of institutions.

Supporting spiritual formation consists of providing repeatable, non-coercive contexts (hospitality, prayer/stillness, moral seriousness, service, truthful conversation, mentorship, and "exit-easy" inquiry) that re-anchor conscience and desire, strengthen the capacity for worship and trust, and produce durable fruits in character and relationships.

The first achievement of such groups is not agreement; it is safety. When encounter is safe, the defenses that late modern culture has learned—against recruitment, against hidden agendas, against identity capture—begin to relax. Only then can a large text be approached as architecture rather than as a test of loyalty or belonging. Only then can Adjuster-illuminated interior commitments take root without being confused with social conformity.

What is most critical given the interregnum context is neither institution-building nor attempts to create an attraction on the horizontal axis, but the creation of a spiritualized social ecology in which persons can become steadily committed to finding and doing the will of God.

In this ecology, book distribution still matters—but it becomes a form of stewardship rather than a campaign. The Urantia Papers are present, available, discoverable, but not wielded. They are offered as a resource to those whose inward questions have ripened enough to receive a large framework without compressing it into a slogan or a badge.

The decisive safeguard is interior: If the person's primary orientation is already shifting toward the will of God under the leading of the Thought Adjuster, the book is less likely to be treated as an identity object; it is more likely to be received as orientation architecture.

Soulcraft then, is fostering ultimacy where it belongs--not in outcomes, not in reach, not in number of clicks, not in brand visibility, but in the living reality of God encountered within the soul and expressed in the spiritual seeker.

The governing intention is the emergence of replicable fruits in lives increasingly committed to finding and doing the will of God. Soulcraft is the upstream cultivation of that commitment—orientation to the Adjuster's leading—without which no downstream stabilization of community, institution, or revelation can remain humane, durable, and free of coercion.

The challenge facing all sincere men and women born of the Spirit is to catalyze the transformation of individuals, the fostering of those new teachers of Jesus' religion who will be exclusively devoted to the spiritual regeneration of humanity. These will be the spirit-born souls who will quickly supply the leadership and inspiration requisite for the social, moral, economic, and political reorganization of the world.

Appendix I: Social Theory and the Urantia Papers on the Interregnum

Contemporary social theory names the sociological symptoms and dynamics of the interregnum; the Urantia Papers provide the anthropological and spiritual architecture that explains why those dynamics occur and how recovery proceeds.

1) Interregnum as cultural-moral disintegration, not institutional underperformance

Urantia Papers references

Paper 99 – "The Social Problems of Religion": Argues that while religion is primarily an inner, personal reality, it exerts decisive social effects by moralizing conscience and re-ordering loyalties—so that durable social renewal depends on spiritually transformed persons rather than on mere institutional control.

Paper 71 – "Development of the State": Distinguishes coercive order from value-based order; durable governance depends on moral development and shared social values, not mere force.

Parallels in sociological studies

Antonio Gramsci: "Interregnum"—the old order is dying while the new cannot yet be born; emphasis on hegemony (cultural-moral coherence) rather than institutional control alone.

Max Weber: Legitimacy/authority typologies (traditional/charismatic/legal-rational); distinction between rule-following and belief in rightfulness.

Émile Durkheim: Anomie and breakdown of shared norms; cohesion as moral integration, not administrative output.

Why it fits: Papers 99 and 71 frame the same structural condition the sociologists diagnose: institutions lose integrative force when moral consensus thins.

2) Obedience without belief: compliance substituting for allegiance

Urantia Papers references

Paper 69 – "Primitive Human Institutions": Customs and regulations can persist mechanically after their moral vitality fades; living values vs inherited forms.

Paper 71 – "Development of the State": Law without ethical interiorization becomes brittle and increasingly coercive.

Parallels in sociological studies

Max Weber (again): Legal-rational authority's dependence on legitimacy; mere legality cannot carry authority when belief erodes.

Institutional sociology / political science (general tradition): Procedural compliance vs normative assent; systems can remain operational while hollowing out motivationally.

James C. Scott (broadly): Formal order without moral buy-in yields "public compliance / private evasion" dynamics.

Why it fits: The Weberian distinction between *rule-following* and *belief-in-rightfulness* parallels The Urantia Papers' distinction between *external regulation* and *internalized moral motivation* almost exactly.

3) Narrative as civilizational glue

Urantia Papers references

Paper 87 – "The Ghost Cults" and Paper 92 – "The Later Evolution of Religion": Religion as integrating horizon—binding generations, organizing meaning, supplying teleology beyond survival.

Paper 99 – "The Social Problems of Religion": Addresses erosion of religious meaning and resulting fragmentation; warns against religion becoming formal/transactional.

Parallels in sociological studies

Peter L. Berger (and allied sociology of religion): "Sacred canopy" / meaning-integration function of shared horizons.

Charles Taylor: "Social imaginary," moral sources, and modernity's reshaping of background conditions of belief.

Alasdair MacIntyre: Virtue depends on tradition-constituted practices and narratives; fragmentation as loss of shared teleology.

Why it fits: Berger, Taylor, and MacIntyre theorize religion's narrative and teleological functions in essentially the same structural register the Urantia Papers describe.

4) Anthropological primacy: persons, conscience, and inward assent

Urantia Papers references

Paper 2 – "The Nature of God" (personality/relationship)

Paper 5 – "God's Relation to the Individual"

Paper 16 – "The Seven Master Spirits" (personality as irreducible cosmic fact)

Paper 110 – "Relation of Adjusters to Individual Mortals"

(These Papers collectively ground value and orientation in personhood, conscience, and inward experience—not collective machinery.)

Parallels in sociological / philosophical studies

Personalist philosophy (broad stream): Personhood irreducible; value rooted in persons and relations; Buber's I–Thou;

Kierkegaard on inwardness/seriousness; Polanyi on tacit commitment/fiduciary knowing: Rational knowledge rests on prior acts of trust in people, testimony, instruments, communities of inquiry.

Hannah Arendt: Loneliness, erosion of a "common world," and susceptibility when shared meaning collapses. Totalitarianism doesn't start mainly with fanatic belief; it starts with people made socially and existentially available for domination.

Why it fits: Urantia Papers anthropology is functionally personalist: value originates in persons; Arendt's loneliness motif parallels the loss of inwardness and moral self-direction.

5) The "vertical axis" framing: ultimacy as ordering horizon

Urantia Papers references

Paper 1 – "The Universal Father": Establishes the ultimate reference point for personality and value.

Paper 111 – "The Adjuster and the Soul": Describes the interior axis of moral/spiritual orientation.

Paper 196 – "The Faith of Jesus": Lived ultimacy—faith as orienting horizon ordering conduct and integrating meaning.

Parallels in sociological / anthropological studies

Religious anthropology / theology of ultimacy (general tradition): Societies require an ultimate reference; when it collapses, it is replaced by functional substitutes.

Contemporary meaning/legitimation theory (general): "Ordering horizons" stabilize valuation and restraint; absent a vertical reference, surrogate ultimacies proliferate.

Why it fits: "Vertical axis" is a contemporary abstraction of what these Papers present as God-consciousness, faith-orientation, and Adjuster-led conscience.

6) The causal ordering: persons → communities → narratives → institutions

Urantia Papers references

Paper 68 – "The Dawn of Civilization": Personal habits/family structures scale into social institutions.

Paper 81 – "Development of Modern Civilization": Cultural advance emerges from individual innovation and moral growth before institutional stabilization.

Paper 100 – "Religion in Human Experience": Spiritual experience at the root of social transformation.

Parallels in sociological studies

Persons → Communities (value-generation becomes social reality)

Durkheim: Durable social order presupposes moral integration in persons—shared norms arise from interiorized obligations before they can stabilize collective life.

George Herbert Mead (symbolic interactionism): The self is formed in interaction; reliable agency and mutual recognition are forged in relational settings before any wider structuration can hold.

Communities → Narratives (shared meaning condenses and becomes transmissible)

Halbwachs: "Collective memory" is generated and maintained in groups; communities supply the social scaffolding that selects, preserves, and repeats what later becomes common story.

Goffman: Communities develop frames that organize experience; these interpretive templates function as proto-stories that make coordinated action possible.

Narratives → Institutions (legitimation and codification of purpose)

Weber: Legitimacy reduces governance costs; when an order is narrated and received as rightful, compliance becomes largely voluntary and administration can remain "thin."

Meyer & Rowan / neo-institutionalism: Institutions adopt publicly available cultural "scripts" (legitimating models) that originate in the social field; narrative models become routinized as formal roles, procedures, and mission.

Institutions consolidate and formalize what is already socially vivified

Berger & Luckmann: institutionalization is the stabilization of already-shared meanings into objective patterns ("this is how we do things"), maintained through legitimation and socialization.

Swidler / Sewell: culture functions as toolkits/schemas in practice; institutions selectively stabilize, amplify, and resource what is already operative in social life rather than generating values ex nihilo.

Why it fits: The Urantia Papers repeatedly emphasize that civilization grows outward from transformed persons; Durkheim/Weber/narrative theory converges on the same ordering in their own idioms.

Influential Works / Related Reading

Arendt, Hannah. *The Origins of Totalitarianism.* New York: Harcourt, Brace and Company, 1951.

Berger, Peter L. *The Sacred Canopy: Elements of a Sociological Theory of Religion.* Garden City, NY: Doubleday, 1967.

Berger, Peter L., and Thomas Luckmann. *The Social Construction of Reality: A Treatise in the Sociology of Knowledge.* Garden City, NY: Anchor Books, 1966.

Blumer, Herbert. *Symbolic Interactionism: Perspective and Method.* Englewood Cliffs, NJ: Prentice-Hall, 1969.

Buber, Martin. *I and Thou.* Translated by Walter Kaufmann. New York: Charles Scribner's Sons, 1970.

Comiskey, Joel. *2000 Years of Small Groups: A History of Cell Ministry in the Church.* Moreno Valley, CA: CCS Publishing, 2014.

Crewdson, Joan O. *Christian Doctrine in the Light of Michael Polanyi's Theory of Personal Knowledge: A Personalist Theology.* Toronto Studies in Theology 66. Lewiston, NY: Edwin Mellen Press, 1994.

Davidsen, Markus Altena, ed. *Narrative and Belief: The Religious Affordance of Supernatural Fiction.* 1st ed. London: Routledge, 2018.

Durkheim, Émile. *The Elementary Forms of Religious Life.* Translated and with an introduction by Karen E. Fields. New York: Free Press, 1995.

Fleming, David L., S.J. *What Is Ignatian Spirituality?* Chicago: Loyola Press, 2008.

Goffman, Erving. *Frame Analysis: An Essay on the Organization of Experience.* New York: Harper & Row, 1974.

Gramsci, Antonio. "'Wave of Materialism' and 'Crisis of Authority.'" In *Selections from the Prison Notebooks of Antonio Gramsci,* edited and translated by Quintin Hoare and Geoffrey Nowell Smith, 275–76. New York: International Publishers, 1971.

Halbwachs, Maurice. *On Collective Memory.* Edited, translated, and with an introduction by Lewis A. Coser. Chicago: University of Chicago Press, 1992.

Henderson, D. Michael. *John Wesley's Class Meeting: A Model for Making Disciples.* Nappanee, IN: Evangel Publishing House, 1997.

Kuhn, Thomas S. *The Structure of Scientific Revolutions.* Chicago: University of Chicago Press, 1962.

MacIntyre, Alasdair. *After Virtue: A Study in Moral Theory.* 3rd ed. Notre Dame, IN: University of Notre Dame Press, 2007.

McAdams, Dan P., Ruthellen Josselson, and Amia Lieblich, eds. *Identity and Story: Creating Self in Narrative.* The Narrative Study of Lives 4. Washington, DC: American Psychological Association, 2006.

Mead, George Herbert. *Mind, Self, and Society: From the Standpoint of a Social Behaviorist.* Edited by Charles W. Morris. Chicago: University of Chicago Press, 1934.

Meyer, John W., and Brian Rowan. "Institutionalized Organizations: Formal Structure as Myth and Ceremony." *American Journal of Sociology* 83, no. 2 (September 1977): 340–363.

Rahner, Karl, S.J. *Foundations of Christian Faith: An Introduction to the Idea of Christianity.* Translated by William V. Dych. New York: Seabury Press, 1978.

Ricoeur, Paul. "Toward a Narrative Theology: Its Necessity, Its Resources, Its Difficulties." In *Figuring the Sacred: Religion, Narrative, and Imagination,* edited by Mark I. Wallace, translated by David Pellauer, 236–248. Minneapolis: Fortress Press, 1995.

Rutler, George William. *A Crisis in Culture: How Secularism Is Becoming a Religion.* Manchester, NH: Sophia Institute Press, 2022.

Spengler, Oswald. *The Decline of the West.* Translated by Charles Francis Atkinson. 2 vols. New York: Alfred A. Knopf, 1926–1928.

Taylor, Charles. *A Secular Age.* Cambridge, MA: Belknap Press of Harvard University Press, 2007.

Tillich, Paul. *Dynamics of Faith.* New York: Harper & Brothers, 1957.

Uversa Press. *The Urantia Book.* New York: Uversa Press, 2015

Weber, Max. *Economy and Society: An Outline of Interpretive Sociology.* Edited by Guenther Roth and Claus Wittich. 2 vols. Berkeley: University of California Press, 1978.

About the Author

David Kantor is a multimedia content creator focused on fostering informed interest in The Urantia Book. A retired software developer and a reader since 1966, his Urantia related work spans the production of feature-length documentary films, books, websites, conference talks, workshops, and technical papers.

With a background in music performance, photography, and professional graphics work, he began producing multiple-projector media performances in the late 1970s. As director of the multimedia division of the Family of God Foundation, he created programs for regional and national Urantia conferences.

He pioneered some of the first websites devoted to The Urantia Book and its history, providing an archive of secondary resources.

As chair of The Urantia Book Fellowship's International Fellowship Committee for many years, he helped support reader conferences across Latin America—including Brazil, Chile, Colombia, Ecuador, Costa Rica, Guatemala, and Mexico—as well as Lithuania and Russia.

In 2012, he founded Urantia Book Films under the auspices of Rocky Mountain Spiritual Fellowship to develop mass-market media on Urantia Book themes. He has produced and directed two feature length films—*Re-Imagining Jesus* (2014), shot on location in Israel and Jordan, and *When Stories Die: Christianity's Crisis and the Urantia Revelation* (2025).

In 2018 he served as an associate editor for the Urantia Press publication, *The Untold Story of Jesus.*

He is the author of *Seed, Flower, Fruit: From the Kingdom of Heaven to the Supreme Being (2025), Beside the Still Waters: 30 Psalms for Readers of The Urantia Book (2025),* and *A Personalist Cosmology: Orientation to the Core Theology of The Urantia Book (2-026).*

David lives in the glorious state of Colorado and is an active member of Rocky Mountain Spiritual Fellowship.

www.ingramcontent.com/pod-product-compliance
Lightning Source LLC
LaVergne TN
LVHW022307200926
848481LV00004B/1238
9798234016461